ISSUES today

A resource for **KS3**

Recycling

Editor: **Claire Owen**

ISSUE 22

Independence
Educational Publishers
Cambridge

First published by Independence
The Studio, High Green, Great Shelford
Cambridge CB22 5EG
England

British Library Cataloguing in Publication Data

Recycling – (Issues Today Series)
I. Owen, Claire II. Series
363.7'282'0941
ISBN 978 1 86168 481 3

Acknowledgements

The publisher is grateful for permission to reproduce the following material.

While every care has been taken to trace and acknowledge copyright, the publisher tenders its apology for any accidental infringement or where copyright has proved untraceable. The publisher would be pleased to come to a suitable arrangement in any such case with the rightful owner.

Chapter One: The Waste Problem

What is rubbish?, © Recycle More, *Wacky waste facts*, © Waste Online, *Waste management*, © rubbish.co.uk, *Wasted food*, © WRAP, *Packaging waste problem*, © Crown copyright is reproduced with the permission of Her Majesty's Stationery Office, *The problem with litter*, © Campaign to Protect Rural England, *Carrier bags*, © Crown copyright is reproduced with the permission of Her Majesty's Stationery Office, *British waste adds to environmental crisis across China*, © Guardian

Chapter Two: Recycling

Waste at home, © Waste Online, *Facts and figures*, © Recycling Guide, *Home composting*, © Recycle Now, *How much of a waster are you?*, © Waste Watch, *Recycling tips*, © Recycle More, *What happens to our recycling?*, © Recycle Now, *Recycle on the go*, © edie, *Why recycling isn't really saving the planet*, © Institute of Economic Affairs/Yorkshire Post, *Consumers oppose charges for plastic bags*, © Ipsos MORI, *Shipping our plastic bottles to China*, © WRAP.

All illustrations, including the cover, are by Don Hatcher.

Design and production, on behalf of Independence, by Hart McLeod Limited, Cambridge.

Printed in Great Britain by MWL Print Group Ltd.

Claire Owen
Cambridge
January, 2009

ISSUES today

Recycling

Contents

Chapter One: The Waste Problem

Chapter Two: Recycling

Issues today A resource for KS3

About Key Stage 3

Key Stage 3 refers to the first three years of secondary schooling, normally years 7, 8 and 9, during which pupils are aged between 11 and 14.

This series is also suitable for Scottish P7, S1 and S2 students.

About Issues Today

Issues Today is a series of resource books on contemporary social issues for Key Stage 3 pupils. It is based on the concept behind the popular *Issues* series for 14- to 18-year-olds, also published by Independence.

Each volume contains information from a variety of sources, including government reports and statistics, newspaper and magazine articles, surveys and polls, academic research and literature from charities and lobby groups. The information has been tailored to an 11 to 14 age group; it has been rewritten and presented in a simple, straightforward format to be accessible to Key Stage 3 pupils.

In addition, each *Issues Today* title features handy tasks and assignments based on the information contained in the book, for use in class, for homework or as a revision aid.

Issues Today can be used as a learning resource in a variety of Key Stage 3 subjects, including English, Science, History, Geography, PSHE, Citizenship, Sex and Relationships Education and Religious Education.

About this book

Recycling is the twenty-second volume in the *Issues Today* series. It looks at waste and recycling in the UK. We all produce waste of some sort and the UK sends more rubbish to landfill than any other country in the European Union. By reducing, reusing and recycling our waste we can save natural resources and help protect the environment. How can we tackle the waste problem and recycle more of our rubbish?

Recycling offers a useful overview of the many issues involved in this topic. However, at the end of each article is a URL for the relevant organisation's website, which can be visited by pupils who want to carry out further research.

Because the information in this book is gathered from a number of different sources, pupils should think about the origin of the text and critically evaluate the information that is presented. Does the source have a particular bias or agenda? Are you being presented with facts or opinions? Do you agree with the writer?

At the end of each chapter there are two pages of activities relating to the articles and issues raised in that chapter. The 'Brainstorm' questions can be done as a group or individually after reading the articles. This should prompt some ideas and lead on to further activities. Some suggestions for such activities are given under the headings 'Oral', 'Moral dilemmas', 'Research', 'Written' and 'Design' that follow the 'Brainstorm' questions.

For more information about *Issues Today* and its sister series, *Issues* (for pupils aged 14 to 18), please visit the Independence website.

www.independence.co.uk

What is rubbish?

RUBBISH IS EVERYTHING THAT YOU THROW AWAY or no longer have a use for. It is anything from an empty crisp packet to a broken toy. Rubbish can be solid, liquid or gas. People sometimes use the word 'waste' when talking about rubbish.

People usually talk about three different sorts of rubbish:

1. **domestic** – rubbish from households.
2. **industrial and commercial** – rubbish from factories, offices, shops and schools.
3. **hazardous** – rubbish which needs to be disposed of in a careful way to prevent pollution, for example, chemicals used to make paint.

Rubbish is made up of different substances and materials:

- **biodegradable rubbish** – breaks down naturally in the environment and eventually disappears. For example, left-over food and paper.
- **non-biodegradable rubbish** – does not break down naturally in the environment. For example, drinks cans and plastic bottles.

What's in your bin?

Household waste is made up of a number of different materials:

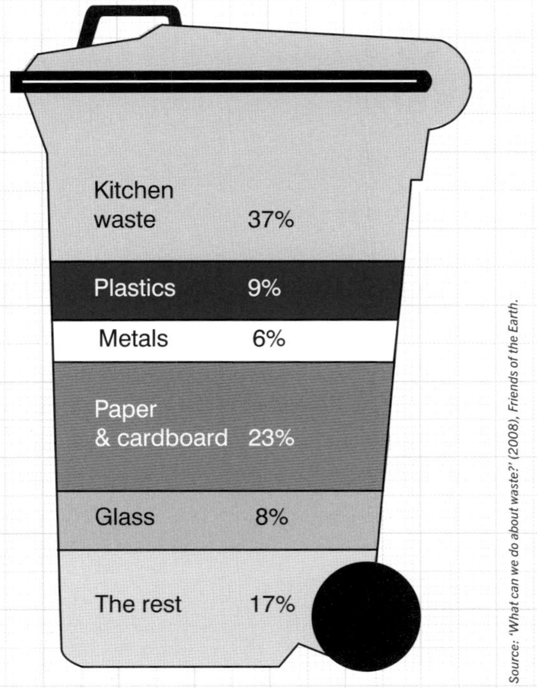

Certain laws make sure that rubbish is disposed of in a responsible way. For example, if chemicals find their way into rivers or streams they can cause harmful pollution, which may kill wildlife. If rubbish from dustbins is allowed to overflow then the streets become filled with litter.

Remember! What is rubbish to you may be very useful to somebody else. Your old toys or clothes that you have grown out of may be passed to a charity shop.

DID YOU KNOW? *In 2005/06 on average each person in the UK produced 511kg of waste, 74 per cent of which was not recycled.*

What is rubbish?

Why do we recycle?

Each household produces about one tonne of rubbish annually, amounting to about 29.1 million tonnes for the UK each year. That's the same weight of around 4.85 million male African Elephants!

The amount of rubbish we throw away is increasing for a number of reasons:

- new packaging materials and technology are being developed
- lifestyle changes: for example a greater reliance on fast and convenience food
- increasing wealth, leading to greater consumption of goods
- increasing population.

When compared to pre-1960s rubbish, today's rubbish contains more products that don't break down when they're put in the ground. Packaging waste makes up about a quarter of all the rubbish you put in your bin. Most of this could be recycled.

We need to increase the amount of rubbish that is recycled, because we cannot carry on burying and burning rubbish forever. This is recognised by the government who are thinking of ways we can reduce the amount of rubbish we produce and increase the amount we reuse and recycle.

Who makes rubbish?

Everybody makes rubbish. Each week the average family in a developed country gets through 4 glass bottles or jars, 13 cans, 3 plastic bottles and 5 kilograms of paper. That means that every day, a total of about 8,000 tonnes of rubbish is thrown away by families in England and Wales – that amount of rubbish is the same weight as 1,600 African elephants!

The amount of rubbish is growing every year. In 2006/07, we threw away 29.1 million tonnes of municipal waste, 1.4 per cent more than the year before. The good news is that more of this is being recycled than ever before. Household recycling and composting rates in England increased to 31% in 2006/7 from 27% in 2005/6. However, this still leaves a lot of rubbish that we have to burn (incinerate) or bury (landfill).

Waste and recycling

Estimated total annual waste arisings, by sector 2004

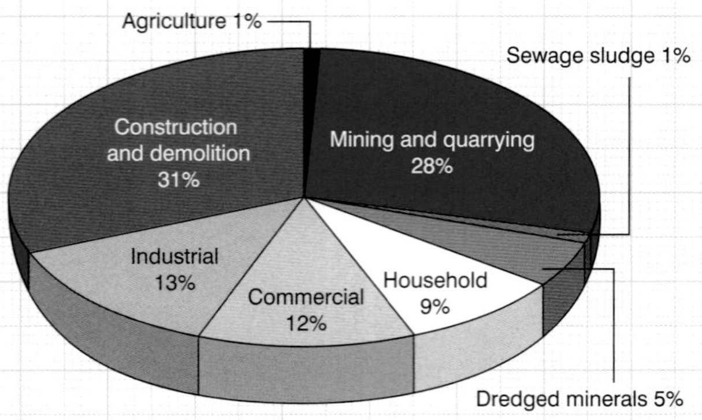

Source: 'Key facts about: Waste and Recycling' (2006), Defra. Crown copyright.

Who produces the most rubbish?

The rubbish from businesses and industries makes up the largest proportion of rubbish collected in the UK. The rubbish comes from places like shops and offices, and industries like manufacturing, building and transport. The pie chart to the left shows where the 335 million tonnes of waste which were produced in the UK in 2004 came from.

www.recycle-more.co.uk

Wacky waste facts

HAVE YOU EVER THOUGHT about how much rubbish you and your family throw away every week? Or why we need to stop throwing so much of it away?

Did you know that...

General garbage

- In 2004 the UK produced about 335 million tonnes of waste. This rate of rubbish production would fill the Albert Hall in London in less than two hours.
- Every year UK households throw away the equivalent of 3 million double-decker buses (almost 30 million tonnes), a queue of which would stretch from London to Sydney (Australia) and back.
- On average, each person in the UK throws away seven times their body weight (about 500kg) in rubbish every year.

Glass

- On average, every family in the UK uses around 330 glass bottles and jars a year. (British Glass)
- It is not known how long glass takes to break down but it is so long that glass made in the Middle East over 3000 years ago can still be found today.
- Recycling two bottles saves enough energy to boil water for five cups of tea.

Paper

- About one-fifth of the contents of household dustbins is made up of paper and card, of which half is newspapers and magazines. This amounts to over 4kg of waste paper per household in the UK each week.

Fantastic plastic

- Every year, an estimated 13 billion plastic bags are given away by supermarkets. This is equivalent to over 290 bags for every person in the UK.
- We produce and use 20 times more plastic today than we did 50 years ago!

Revolting rubbish

- Babies' nappies makes up about 2% of the average household rubbish. This is the same weight as nearly 70,000 double decker buses every year. If lined up front to end, the buses would stretch from London to Edinburgh.

Oil

- 1 litre of oil can pollute 1 million litres of fresh drinking water (Scottish Oil Care Campaign).
- Waste oil from nearly 3 million car oil changes in Britain is not collected. If collected properly, this could meet the annual energy needs of 1.5 million people. (Scottish Oil Care Campaign)

Packaging

- In 2001 UK households produced the equivalent weight of 245 jumbo jets per week in packaging waste.
- Every year each person produces four times as much packaging waste as their luggage allowance on a jumbo jet.

9 June 2008

www.wasteonline.org.uk

Waste management

WE HAVE BECOME CASUAL ABOUT WASTE in our throwaway world, and leave it to the authorities to deal with it. But the pressure on landfill sites demands new solutions, and this means that, increasingly, environmental responsibility for waste disposal will be redirected back towards consumers and manufacturers.

The amount of rubbish we produce has grown rapidly over the past 100 years, with the unstoppable rise of consumer goods, packaging and plastics.

Waste collection

In Britain waste collection is the responsibility of the local authorities and the government. Efficient and effective waste management is an essential part of civilised society. Most people have a fair idea of what they can and cannot put in their rubbish bins; it is the duty of the local authority to keep reminding them of this.

Hazardous waste

Some kinds of waste must be kept apart from the usual stream of waste disposal. Hazardous waste is any kind of waste that could be a serious danger to human beings or to other living organisms. It includes such things as chemicals, medical waste, fuel, paint, asbestos, gas canisters, pesticides, car batteries. By far the biggest amount of hazardous waste is produced by industry – 5 million tonnes of the stuff every year. But individual homeowners also produce their share: if you are in any doubt about how to dispose of such items, consult your local authority.

Landfill

What happens to all this waste after it leaves your home? The most common destination for rubbish worldwide is a landfill site. The rubbish is tipped into a big hole in the ground – usually a hole created by quarrying or mining. In Britain about 70 per cent of waste goes to landfill sites.

The decomposition that takes place in a landfill produces flammable, toxic gases like methane, an environmentally damaging greenhouse gas. However, this can be flared off through pipes, or it can be piped to a plant and burnt as a fuel to generate electricity. Once full, landfill sites can be covered over and sealed, and eventually the land can be used for airports, golf courses, dry ski-slopes, and so on.

Landfill is a relatively cheap option. The trouble is that we are quickly running out of suitable sites. Existing sites are filling up fast, and plans for new sites are opposed by people living in the area.

From an environmental point of view, landfills are a poor option. The EU wants the UK's use of landfills to be cut to half its 1995 levels by 2013, and to 35 per cent by 2020 – a tough agenda. The key to success in this strategy lies in dramatically reducing the amount of waste produced in the first place, largely through recycling.

> ***"Efficient and effective waste management is an essential part of civilised society."***

Waste management

Incineration

So, if you cannot dump rubbish on the land, why not burn it? Incinerators reduce 85 per cent of rubbish to fine ash. This is what already happens to about 10 per cent of Britain's rubbish. We can also use the rubbish as a fuel to produce electricity. The target is to increase the amount of rubbish that is incinerated to 27 per cent by 2020.

Unfortunately, rubbish incinerators produce a large volume of toxic waste, in the form of gas and solids, causing air pollution. In recent years, great technical advances have been made in the process of cleaning ('scrubbing') gas emissions from incinerators, but these add considerably to the costs, and some remaining toxic pollutants still escape. Another criticism is that incineration destroys large amounts of potentially recyclable materials, so the energy loss is substantial. And it is difficult to build new incinerators, because – as with landfills – no one wants to live near to one.

The 'Three Rs' of rubbish

The search for a solution to waste disposal has produced the so-called 'waste hierarchy', or the 'Three Rs' of rubbish: Reduce, Re-use, Recycle. Through this policy we edge towards the goal of producing no rubbish at all: a world of zero-waste. Everything could, in principle, be re-used or recycled.

Municipal waste managment

UK municipal waste management: 2005/06

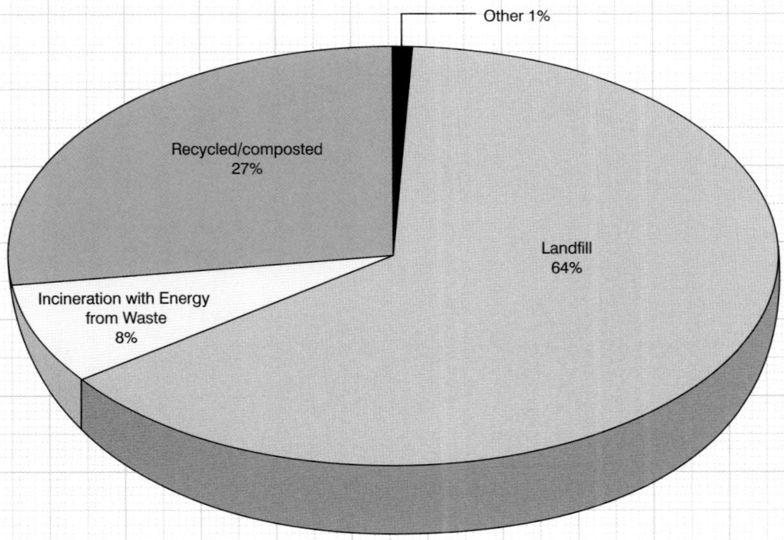

Source: Department for Environment, Food and Rural Affairs (Defra); Environment and Heritage Service, Northern Ireland (EHSNI); Scottish Environment Protection Agency (SEPA); Welsh Assembly Government (WAG).

Mini glossary

- ***organisms*** – *living things*
- ***pesticides*** – *chemicals used to kill pests such as insects or weeds*
- ***quarrying*** – *extracting stone from a mine*
- ***flammable*** – *material that catches fire easily*
- ***decomposition*** – *breakdown of organic material*
- ***toxic*** – *poisonous*

www.rubbish.co.uk

Wasted food

THE COST OF NEEDLESSLY WASTED FOOD to UK households is £10 billion a year, £2 billion higher than previously estimated according to research published by WRAP.

The research reveals that the average household throws out £420 of good food a year. For the average family with children it's higher at £610 – money which could have helped pay household bills.

Thrown out

Researchers found that more than half the good food thrown out, worth £6 billion a year, is bought and simply left unused or untouched. For example, each day in the UK we throw away:

- 1.3 million unopened yoghurt pots
- 5,500 whole chickens
- 440,000 ready meals.

The study revealed that £1 billion worth of wasted food is still 'in date'. It costs local authorities £1 billion a year to dispose of food waste.

Environment Minister Joan Ruddock said:

'These findings are staggering in their own right, but at a time when global food shortages are in the headlines this kind of wastefulness becomes even more shocking.

'This is costing consumers three times over. Not only do they pay hard-earned money for food they don't eat, there is also the cost of dealing with the waste this creates. And there are climate change costs to all of us of growing, processing, packaging, transporting, and refrigerating food that only ends up in the bin.'

Key facts from the Love Food Hate Waste campaign

- In the UK we are throwing away one third of the food we buy. That's like one in three bagfuls of food shopping going straight in the bin.
- We throw away 6.7 million tonnes of food each year in the UK, when most of this food could have been eaten. (It's not just peelings and bones –it's good food). That much food waste would fill Wembley Stadium eight times over!
- In terms of environmental impact – producing, storing and getting the food to our homes uses a lot of energy. The carbon impact of food waste is enormous. Tackling it would provide the same carbon benefit as taking one in five cars off of UK roads.
- Most of the wasted food reaches landfill sites where it releases methane, a powerful greenhouse gas.
- We throw food out for two main reasons: food gets forgotten and is left unused; we serve up too much and don't use leftovers.

8 May 2008

www.wrap.org.uk

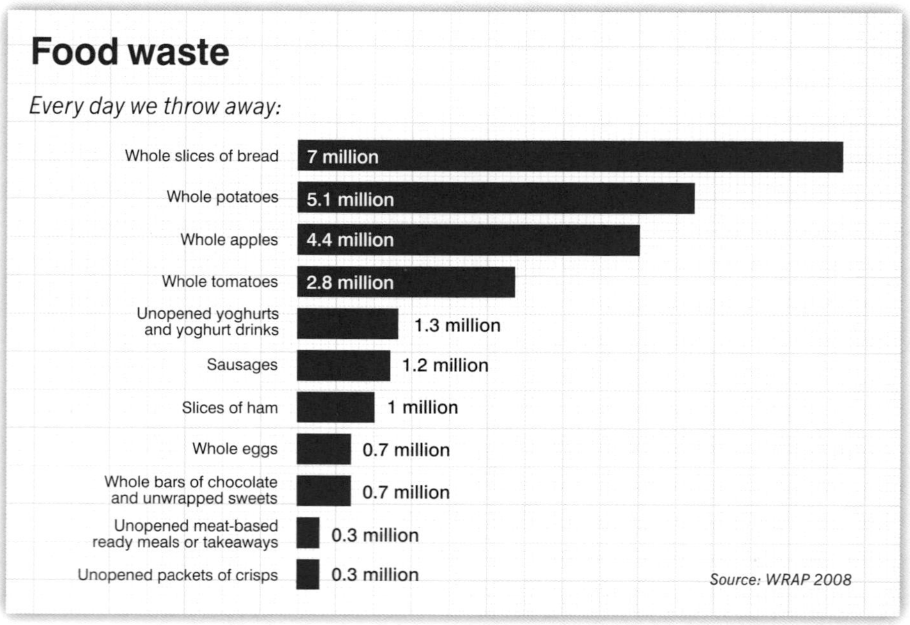

For more information on the Love Food Hate Waste campaign, visit: www.lovefoodhatewaste.com

Packaging waste problem

THE LOCAL GOVERNMENT ASSOCIATION has published its second investigation into the weight of retailer packaging and how much of it can be recycled. It shows that up to 38 per cent of packaging in a regular household shopping basket cannot be recycled. The first survey, published in October 2007, put that figure at 40 per cent.

Less packaging

The British Market Research Bureau (BMRB) was commissioned by the LGA to buy a range of common food items from eight retailers. Analysis of the purchases found that local retailers and market traders produced less packaging and more that could be recycled than the larger supermarkets.

Most retailers have a lower weight of packaging than in the first survey – the average weight of packaging has reduced by five per cent – though the proportion that is recyclable has changed little.

DID YOU KNOW ?

In 2007 the Waste Electrical and Electronic Equipment (WEEE) directive came into force, making manufacturers responsible for financing the collection, recovery and disposal of electronic equipment.

Cllr Paul Bettison, Chairman of the LGA Environment Board, said:

'The days of the cling film coconut must come to an end. We all have a responsibility to reduce the amount of waste being thrown into landfill, which is damaging the environment and contributing to climate change.

'Families will be pleased to see that more packaging in their shopping baskets can now be recycled. However, this survey shows there is still a lot further to go.'

www.lga.gov.uk

Recycling rates

Recycling rates have increased to 33 per cent in England as local people do their bit to reduce the amount of waste being thrown into landfill – but this figure must continue to rise. Councils have to pay £32 in tax for every tonne of rubbish that is sent to landfill. That figure will rise to £48 a tonne by 2010. In addition, from 2010 councils face EU fines of £150 for every tonne that is dumped, which could cost an estimated £200 million by 2013.

The LGA is calling on the Government to make retailers and producers responsible for funding the collection of packaging so they are incentivised to cut back on it in the first place. Other countries in the European Union operate a system where companies contribute towards recycling services and household collections.

29 May 2008

Mini glossary

incentivised *– motivated*

The problem with litter

Litter

- It costs taxpayers more than half a billion pounds each year to clear the streets of England, and that doesn't include parks or other public spaces.
- It is illegal to drop litter and you can be fined up to £80 on the spot if you're caught littering.
- Nearly half the population (48%) admit to dropping litter.
- Seven out of ten items of litter are food related.

The effects of litter

- Over 69,000 animals were killed or injured by litter last year in Britain.
- Litter, such as cigarette butts, plastic bags and other plastics, harms animal and marine life in a variety of ways such as water pollution, when these items are mistaken for food and when creatures get caught up in plastics and get strangled. It is estimated that every year over 1 million seabirds and 100,000 turtles and sea mammals die of litter related causes.
- Littered items are a lost resource. When things that could otherwise be recycled, like PET, glass bottles and paper, are littered, they do not end up in the recycling stream.

How much litter is dropped?

- The amount of litter dropped yearly in the UK has increased by 500% since the 1960s.
- An estimated 122 tons of cigarette butts and cigarette-related litter is dropped every day across the UK.
- Over 373,000 pieces of litter were found on beaches in the UK in 2006.

Biodegradability

Any type of litter takes a long time to disappear naturally, so whatever the material the right thing to do is not to drop it in the first place. Degradability depends on climate and circumstances, but under unfavourable conditions estimated time spans are:

- orange peel and banana skins – up to 2 years
- cigarette butts – up to 2 years
- plastic bags –10-20 years
- tin cans – 50 years
- aluminum cans – 80-100 years
- plastic bottles – indefinitely
- glass – indefinitely.

Facts about fly-tipping

Fly-tipping is the common term used to describe waste illegally left on land. A fly-tip may be anything from a single bin bag up to thousands of tonnes of construction and demolition waste. Fly-tipping poses a threat to humans and wildlife, damages our environment and spoils our enjoyment of our towns and countryside.

- A new incident of fly-tipping occurs every 12 seconds at a cost to the public of around £72 a minute.
- Black bags full of domestic rubbish account for 63% of all fly-tipping.
- Fly-tipping can result in fines of up to £20,000 and/or six months imprisonment.

www.cpre.org.uk

Carrier bags

What's the problem?

The 13 billion carrier bags which are distributed in the UK each year make up only a small part of domestic waste. However, they can be a particularly visible form of litter when thrown away irresponsibly, and their disposable, easily-replaced nature make them a symptom of our 'throwaway society' – and public opinion recognises this.

The Government's Waste Strategy for England 2007 set out the Government's aim that free, single-use carrier bags (both paper and plastic) should become a thing of the past. Changing our behaviour in this way is vital if we are to achieve a more sustainable lifestyle.

What have we done so far?

On 28 February 2007, a voluntary agreement was announced with UK retailers to reduce the overall environmental impact of carrier bags by 25% by the end of 2008. 22 major retailers and six trade associations signed up to the agreement.

We welcome the action taken by retailers to cut the environmental impacts of their carrier bags. However, it is clear that the current voluntary arrangement has not resulted in a big enough reduction in the number of bags given out, and will not get us close to achieving our goal of phasing out free, single-use carrier bags.

Charges for single-use carrier bags

The Government has, therefore, decided that stronger action needs to be taken. It was announced in the Budget, on 12 March 2008, that the Government will bring forward legislation in the Climate Change Bill to enable us to require retailers to place a minimum charge on single-use carrier bags, if enough progress is not made on a voluntary basis. These powers will come into force in 2009.

Key facts and figures

- More than 13 billion carrier bags are distributed in the UK every year.

  These account for approximately:
  - 0.3% of household waste
  - 0.1 – 1% of visible litter in the UK
  - 2% of total litter on UK beaches.

- Limiting the use of disposable bags is an important way in which each of us can take action, as each adult receives on average nearly 300 every year.

- 88% of shoppers currently put all their shopping into free carrier bags. On average shoppers take 3-4 bags at every shopping trip.

- 45% of shoppers claim to have bought a Bag for Life but only 12% use one regularly.

12 March 2008

DID YOU KNOW?

Most carrier bags are made from polyethylene, a type of non-biodegradable plastic which could take hundreds of years to break down.

Mini glossary

phasing out *– gradually stopping the use of*

www.defra.gov.uk

British waste adds to environmental crisis across China

British high street waste is fouling streams and ditches in China despite promises of an environmental crackdown by both governments.

Mai village in Guangdong province, southern China, suffers from a Made-in-Britain eyesore: Tesco and Argos plastic bags choke the waterways, snag on tree branches and contribute to a rotting stench during floods and hot weather.

It is a side-effect of globalisation. Many of these products were manufactured in China, shipped to the UK for use and sent 5,000 miles back for disposal. China exported £12.6bn worth of manufactured goods to the UK last year and received an estimated 1.9m tonnes of rubbish in return. Under EU regulations member countries are not allowed to dump garbage overseas, but are permitted to send sorted waste for recycling.

Environmentalists say this is irresponsible because much of the recycling is carried out in poorly-regulated communities, where health risks and pollution worries are a low priority.

Guangdong is scattered with scavenging centres. In Guiyu and Qingyuan small family-run businesses chop up and melt down toxic plastics and metals from discarded computers, printers and mobile phones. In Nanhai and Shunde factories deal with mounds of plastic bags and bottles. About 20% of the waste comes from overseas according to local sources.

Importing waste

A series of exposés in the domestic and foreign media prompted the government to crack down on the business earlier this year. Guangdong's provincial government banned unlicensed businesses and individuals from importing plastic waste and suspended operations at factories that failed to meet environmental standards.

Last month factories in the most notorious district, the Lianjiao area of Nanhai, were shut down. But most firms simply relocated. In nearby Shenzhen and Shunde businessmen were still reprocessing carrier bags and other waste from the UK.

At 'plastic street' in Mai village, dozens of small plastic recycling firms line the road. Most of the work is done by migrant workers who are paid about £50 a month. Thousands of plastic carrier bags were being blown into ditches and waterways, creating an eyesore and a bad smell.

Greenpeace believes that wealthier countries should deal with their waste problems at home, rather than exporting them to developing countries, which have to pay the environmental costs.

'If we can stop the waste trade I am sure it will lead to more sustainable development around the world,' said Kevin May, toxics campaign manager at Greenpeace's office in Beijing.

By Jonathan Watts , 31 March 2007

Mini glossary

globalisation – *growing communication and trade between countries*

exposé – *a report that uncovers wrongdoing*

notorious – *well-known for something bad*

relocated – *moved to a new place*

exporting – *selling to another country*

Activities

Brainstorm

Brainstorm to find out what you know about waste.

1. What is waste?

...

...

2. What is landfill?

...

...

3. What is incineration?

...

...

Oral activities

4. We live in a 'throwaway society' and the amount of rubbish we throw away is increasing. In a group, come up with some possible reasons for this increase in waste production.

NOTES ...

...

...

5. Prepare a short talk on littering and fly-tipping in the UK. Use the article *The problem with litter* on page 8 as a starting point.

NOTES ...

...

...

Moral dilemmas

6. You have seen a new mobile phone you really want to buy, although your current phone is fairly new and still works fine. You love to shop and keep up with the latest trends, but you don't want to create unnecessary waste. What do you do?

7. You see someone dropping litter in your area. You think about confronting them, but they are older than you and you don't want to risk getting in a fight. What should you do?

Activities

Research activities

8. Keep a diary of all the waste you produce over one week. Remember to include the waste you produce at school and at home. How much of your waste is thrown away and how much is recycled? Think about how you could cut down on the amount of waste you produce.

9. Visit the WRAP website at www.wrap.org.uk. What are its main aims? Summarise the priority areas the organisation focuses on.

NOTES..........

Written activities

Complete the following activities in your exercise books or on a sheet of paper.

10. Read the article *Waste management* on pages 4–5. Write two lists summarising the main advantages and disadvantages of landfilling and incineration.

11. Write an article suitable for a teen magazine to raise awareness of the waste problem and encourage young people to recycle.

NOTES..........

Design activities

12. Design a reusable 'bag for life' which will be sold in shops and supermarkets to encourage shoppers not to use free plastic bags. Include a catchy slogan on the bag to get your message across.

Waste at home

WE ALL PRODUCE WASTE of some sort, whether it is an empty drinks can or grass clippings from the garden.

So where does it all go?

A total of:

- 72% of municipal waste is landfilled – which means it's buried in the ground.
- 9% is incinerated – which means it's burnt. This is also called energy from waste.

Dealing with our rubbish in this way is not an ideal solution. When we bury or burn our rubbish we are losing valuable natural resources and wasting the energy, water and transport costs used in its production. Landfilling and incineration can harm the environment if they are not properly managed. Many landfill sites are nearly full and we are quickly running out of suitable land, close to where the rubbish is produced, for new sites. In any case, these sites are often unwelcome neighbours – we keep producing the rubbish, but we don't want it disposed of near to where we live.

Mini glossary

consumer *– someone who buys and uses goods and services*

sustainable *– using resources in such a way that our present needs are met without damaging the needs of future generations*

The alternative?

The 3Rs – reduce, reuse, recycle.

Reducing

Every year the amount of rubbish we produce increases and this leads to increased costs for society – both financial and environmental. The majority of the resources that we use to make things – only to throw them away – can't be replaced. As consumers, we can reverse this trend – buy only the right amount of what we really need, choose products with less packaging, and buy from producers using sustainable practices.

Reusing

We can cut down on the amount of rubbish we have to get rid of by reusing our materials. Computers, furniture, clothing – so many items can be reused. Setting the printer to print on both sides of a sheet of paper, repairing our broken appliances and shoes or finding a charity that will make use of them – we help ourselves and others, and delay the point at which materials become waste.

Recycling

Putting materials aside for recycling helps in many ways: we send less rubbish to landfill or incineration, and we save valuable materials and energy – for example, plastic bottles can be converted into fleeces and garden furniture, whilst recycling aluminium cans saves 95% of the energy used in making a new can. New technologies are furthering our ability to recycle what was previously our waste and turn it back into the resources that we need.

Buy recycled

Support those industries that use recycled materials by buying recycled products. Remember that it's not enough just to recycle, buying recycled ensures that the materials you send for recycling are actually used again.

9 June 2008

www.wasteonline.org.uk

Facts and figures

Aluminium

- 24 million tonnes of aluminium are produced each year, 51,000 tonnes of which ends up as packaging in the UK.
- If all cans in the UK were recycled, we would need 14 million fewer dustbins.
- £36 million worth of aluminium is thrown away each year.
- Aluminium cans can be recycled and ready to use in just six weeks.
- Recycling aluminium produces 95% fewer greenhouse gas emissions than making it from raw materials.

DID YOU KNOW?

- *Up to 60% of the rubbish that ends up in the dustbin could be recycled.*
- *The unreleased energy contained in the average dustbin each year could power a television for 5,000 hours.*
- *On average, 16% of the money you spend on a product pays for the packaging, which ultimately ends up as rubbish.*
- *Up to 80% of a vehicle can be recycled.*
- *9 out of 10 people would recycle more if it were made easier.*
- *As much as 30% of waste in the average dustbin could be composted.*

Recycling is an excellent way of saving energy and protecting the environment

Glass

- Each UK family uses an average of 500 glass bottles and jars annually.
- The largest glass furnace produces over 1 million glass bottles and jars per day.
- Glass is 100% recyclable and can be used again and again.
- Glass that is thrown away and ends up in landfills will never decompose.
- One recycled glass bottle would save enough energy to power a computer for 25 minutes.

Plastic

- 275,000 tonnes of plastic are used each year in the UK, that's about 15 million bottles per day.
- Most families throw away about 40kg of plastic per year, which could otherwise be recycled.
- The use of plastic in Western Europe is growing about 4% each year.
- Plastic can take up to 500 years to decompose.
- One recycled plastic bottle would save enough energy to power a 60-watt light bulb for three hours.

Paper

- Recycled paper produces 73% less air pollution than if it was made from raw materials.
- 12.5 million tonnes of paper and cardboard are used annually in the UK.
- The average person in the UK gets through 38kg of newspapers per year.
- It takes 24 trees to make one ton of newspaper.
- 70% less energy is required to recycle paper compared with making it from raw materials.

www.recycling-guide.org.uk

The above information is reprinted with kind permission from Recycling Guide.

Home composting

COMPOSTING IS NATURE'S WAY OF RECYCLING and helps to reduce the amount of waste we put out for the bin men. The following easy guide to home composting will give you all the information needed to get the best out of your bin. Now let's get composting!

STEP 1 Placing your bin

It's best to put your bin on a level, well-drained spot. This allows excess water to drain out and makes it easier for helpful creatures such as worms to get in and get working on breaking down the contents. Placing your bin in a partly sunny spot can help speed up the composting process.

STEP 2 Put these in

Like any recipe, your compost needs the right ingredients to make it work. Good things you can compost include vegetable peelings, fruit waste, teabags, plant prunings and grass cuttings. These are called 'greens.' Greens are quick to rot and they provide important nitrogen and moisture. Other things you can compost include cardboard egg boxes, scrunched up paper and fallen leaves. These are called 'browns' and are slower to rot. They provide fibre and carbon and also allow important air pockets to form in the mixture.

STEP 3 Keep these out

Certain things should never be placed in your bin. Cooked vegetables, meat, dairy products, diseased plants, dog poo or cat litter, or baby's nappies should not be put into the compost bin. Putting these in your bin can encourage unwanted pests and can also create odour. Remember that plastics, glass and metals are not suitable for composting and should be recycled separately.

STEP 4 Making good compost

The key to good compost lies in getting the mix right. You need to keep your greens and browns properly balanced. If your compost is too wet, add more browns. If it's too dry, add some greens. Making sure there is enough air in the mixture is also important. After approximately 9–12 months your finished compost will be ready.

STEP 5 Using your compost

Finished compost is a dark brown, almost black soil-like layer that you'll find at the bottom of your bin. It has a spongy texture and is rich in nutrients. Spreading the finished compost into your flowerbeds greatly improves soil quality by helping it retain moisture and suppressing weeds. Composting is the easiest way to make your garden grow more beautifully.

www.recyclenow.com

recyclenow

How much of a waster are you?

FIND OUT HOW GOOD YOU ARE at looking after the Earth's resources by trying this 'lifestyle challenge' based on a typical school day.

1 *You want to take a drink to school. Do you:*

- ☐ a) buy a can of drink and throw it away?
- ☐ b) take a plastic flask which you can use again and again?
- ☐ c) buy a can of drink and recycle the can?
- ☐ d) refill the plastic bottle that you bought yesterday?

2 *You have a lot of things to carry. Do you:*

- ☐ a) use the school bag which you use every day?
- ☐ b) use a supermarket carrier bag which you'll use today and then throw away?
- ☐ c) use a plastic carrier bag and take it to the supermarket for recycling?
- ☐ d) use the same carrier bag that you used yesterday?

3 *In your first lesson you have to write a story. You make a mistake in the first sentence. Do you:*

- ☐ a) throw the paper in the bin and get a new piece?
- ☐ b) put the paper in the scrap drawer for someone else to use?
- ☐ c) put the paper in the box from which paper is taken for recycling?
- ☐ d) rub out the mistake and carry on?

4 *In maths, your class is given a worksheet to do. The teacher gives you one sheet each and it's only printed on one side. Do you:*

- ☐ a) suggest that they could print it on both sides and just give you one sheet between two?
- ☐ b) put the sheet in the recycling box when you've finished with it?
- ☐ c) give the sheet back to your teacher when you've finished with it so that it can be used again next year?
- ☐ d) keep your mouth shut and get on with the work?

5 *You usually take a packed lunch to school. Do you bring:*

- ☐ a) a resealable plastic box with no other wrapping inside it?
- ☐ b) sandwiches and cakes from the supermarket, individually wrapped in plastic?
- ☐ c) cake wrapped in aluminium foil which can be used again the next day?
- ☐ d) sandwiches wrapped in aluminium foil which you will take home for recycling?

6 *You notice that your school is having a delivery of new paper. Do you:*

- ☐ a) ask whether the head teacher has ordered recycled paper?
- ☐ b) suggest ways in which your school could use less paper in the first place?
- ☐ c) check that everyone in your school is using paper on both sides?
- ☐ d) say nothing – it's none of your business?

How much of a waster are you?

7 *In the afternoon it's design and technology and you're making a model. Do you design your model so that it can be made:*

- ☐ a) only from new sheets of cardboard provided by school?
- ☐ b) from offcuts of cardboard, parts of which have already been used for something else?
- ☐ c) from cereal packets and other cardboard that you've brought from home?
- ☐ d) from a construction kit (e.g. K'nex or Lego) that can be used over and over again?

8 *Oh dear. You've still got some homework to do! Fortunately you can find out a lot of the answers using the internet. Do you:*

- ☐ a) find lots of pages of information and print off all of them?
- ☐ b) only print off those pages that look really useful?
- ☐ c) make notes on the information you need?
- ☐ d) store the most useful pages in your 'Favourites' so that you can look at them offline?

1–20

Everyone has to start somewhere! Take a look at the ideas in the info zone at www.recyclezone.org.uk and pick one easy way to reduce, reuse and recycle. Once you've got that going, try something else and gradually build up your 3Rs power.

21–30

You've made a great start and are well on the way to being a '3Rs Champion'. Take a look at the ideas in the info zone at www.recyclezone.org.uk and pick some more ways to reduce, reuse and recycle to add to those that you already do.

Scores

1. a) 0, b) 5, c) 3, d) 4
2. a) 5, b) 0, c) 3, d) 4
3. a) 0, b) 4, c) 3, d) 5
4. a) 5, b) 3, c) 4, d) 0
5. a) 5, b) 0, c) 4, d) 3
6. a) 3, b) 5, c) 4, d) 0
7. a) 0, b) 4, c) 4, d) 5
8. a) 0, b) 3, c) 3, d) 5

31–40

If you've been honest in your answers, you are already a '3Rs champion'. But don't stop there! Try to reduce or reuse even more than you recycle and make sure that your friends and family know how to be '3Rs Champions' as well. There are plenty of ideas to show them in the info zone at www.recyclezone.org.uk.

www.recyclezone.org.uk

Recycling tips

Batteries

- some councils and garages provide facilities for recycling rechargeable batteries and lead acid car batteries
- rechargeable batteries should be returned to the manufacturer where possible for disposal.

Advice: buy rechargeable batteries, or appliances which use mains electricity.

CDs and DVDs

- you can post your old CDs, DVDs, CDRs, VHS and cassette tapes to The Laundry for recycling
- Polymer Reprocessors recycle all CDs and also the packaging around the CDs.

Advice: reuse your old CDs and DVDs by taking them to a charity shop.

Chemicals, paint and oils

- chemicals are used every day in the home and garden
- DO NOT pour chemicals and oil down drains as they can pollute rivers
- used engine oil can be recycled at most civic amenity sites
- old or leftover paints can be used by community groups such as Community RePaint
- plastic bottles which have contained household cleaners can also be recycled.

Advice: use environmentally friendly chemicals, which are stocked in most DIY stores.

Electronics

- some retailers take back old electrical items when delivering a new one
- if your item still works safely, you could sell it
- some charity shops will accept small electrical items
- there are now lots of reuse networks where you can offer items for others to reuse
- mobile phones can be recycled through phone retailers and charities
- If your old ipod is broken and no longer under warranty you can have it repaired or sell it to: www.ukipodrepairs.com

Advice: try to repair broken items rather than throw them away. Buy durable items with long life cycles.

Plastic

- there are over 50 different types of plastics
- if separate bins are provided it will usually be for:

  HDPE – opaque bottles e.g. detergent bottles

  PVC – transparent bottles, an obvious seam running across the base e.g. mineral water bottles

  PET – transparent bottles, a hard moulded spot in the centre of the base e.g. fizzy drink bottles
- some supermarkets have collection points for recycling carrier bags.

Advice: reuse bags or use a long-life carrier bag. Buy in bulk to reduce packaging.

www.recycle-more.co.uk

Recycling tips

Glass

- bottles and jars are usually separated by colour: brown, clear, and green
- place glass in the correct colour bin
- recycle your blue glass with your green glass
- wash out bottles and jars, remove caps and corks before recycling
- light bulbs, Pyrex-type dishes, windowpanes etc. should not be put in glass banks.

Advice: reuse your jars for storage. Most supermarkets have glass banks where you can recycle alongside your weekly shop!

Medical waste

- dispose of medicines following either your doctor's or the manufacturer's instructions
- care should be taken when disposing of needles and syringes
- glass bottles and jars that have contained medicines can be recycled when they are empty.

Advice: if you find a syringe, use your common sense. If you can safely pick it up, then place it in a safe container and take it to the local police station.

Textiles

- old clothes, bedding, curtains, and blankets can be recycled on any high street at charity shops, but only donate clean usable items!

Advice: use any unrecyclable textiles as cloths around your home.

Metals

- usually separated into: aluminium (drinks cans) – non-magnetic, and steel (food tins) – magnetic. Aerosols can be made from either
- test by using a magnet
- wash and squash cans before recycling. Only recycle clean aluminium foil
- only put empty aerosols in recycling schemes and never pierce or crush aerosols, even when they are empty
- crisp wrappers (metallised plastic film) cannot be recycled. Metallised plastic springs back when scrunched.

Advice: contact Alupro for details of their Cash for Cans scheme and BAMA for any aerosol queries.

Paper and cardboard

- paper collection is usually separated into: newspapers, magazines, cardboard, and phone directories
- unless specified, do not recycle catalogues, directories or envelopes which are gummed or glued together
- juice and milk cartons cannot be recycled with ordinary paper as they are made up of several materials.

Advice: if you read newspapers, please recycle them after use. Alternatively, read news online. Set your printer to print double sided, buy recycled paper.

Furniture

- local charity shops, schools, and community groups can sometimes use unwanted items
- please note that unwanted sofas and chairs must have the kite mark to prove they meet British safety standards
- make sure all furniture is clean and in good repair before you donate it.

Advice: old furniture may be very useful to someone else! Donate unwanted items where possible.

What happens to our recycling?

EVERYTHING GOES TO LANDFILL, doesn't it? Well, actually, no. In the UK more than 95% of the recycling collected is recycled. The remaining 5% is often not recycled because the wrong materials have been put out for collection or are too dirty to be reprocessed.

So where does waste go? If it is not sorted into separate materials by your refuse collectors, your recycling will go to a materials recycling facility to be sorted. The sorted materials are sold to reprocessing companies who turn them back into raw materials; these are then used to make new products.

Here's a peek into the world of waste recycling

Aluminium

Aluminium cans are shredded, melted down, and the molten aluminium poured into moulds to make ingots. These are then sold to companies who make new products such as car and plane parts, or maybe the can containing your next drink!

Glass

Glass is crushed and added to the mix of raw materials that make up new glass containers. The materials are melted in a furnace, and then moulded or blown to make new bottles and jars. Glass is also used to make unusual stuff – for example, it's an ingredient used to make new bricks. The UK currently recycles around 50% of container glass (like bottles and jars). That's doubled over the last 5 years.

Plastic

Sorted plastics have a wide variety of uses and can be shredded, washed, melted and moulded into new products such as new bottles, garden furniture or fleece jackets. It takes about 25 2-litre drinks bottles to make one adult size fleece jacket.

Paper

When paper gets to the recycled paper mill it is added to water and turned into pulp. It is screened, cleaned and where required, de-inked until it is suitable for making new paper products such as newsprint, cardboard, packaging, tissue and office items. It can take just seven days for a newspaper to go through the recycling process and be transformed into recycled newsprint. All the newsprint (the paper for newspapers) manufactured in the UK is now made from 100% recycled paper.

Steel

Steel is a brilliant product to recycle, as it can be reprocessed again and again. Steel cans are melted down in a furnace and combined with other raw materials like molten iron. The hot steel is then cast into solid slabs which can be rolled into foil to make new cans.

www.recyclenow.com

recyclenow

Recycle on the go

APPARENTLY HAPPY WITH PROGRESS on the home front, the Government now hopes to encourage us to embrace recycling while we're out and about.

Launching the Recycle on the Go pilot scheme in London's Hyde Park on Monday, Environment Minister Joan Ruddock told edie that the idea was to provide recycling bins in convenient locations to give people the opportunity to do the right thing.

'We're saying, "look, we know you're not going to take those cans and bottles home with you but now you can recycle them rather than just throwing them in the bin",' she said.

> **"We know people are prepared to recycle and this is about giving them the opportunity to do so."**

'We know people are prepared to recycle and this is about giving them the opportunity to do so.'

She said that previously litter collected in park bins would go to landfill as it was inefficient to try to sort it for recycling.

'Normal bin waste is usually highly contaminated so it is difficult to recycle,' she said.

She acknowledged there was likely to be a relatively high level of contamination initially with these bins, as there had been with kerbside recycling schemes when they were first rolled out, but she said people had learned how to sort their waste at home and she was confident they could do so while out and about too.

She accepted that two bins in one park were unlikely to have a significant impact on general recycling rates, but argued that it was better to start with a pilot scheme to assess the public appetite for recycling while they're on the move, before rolling out a wider scheme.

By Sam Bond, 3 June 2008

DID YOU KNOW? *The amount of household waste recycled or composted in the UK has increased from 6 per cent in 1995/96 to 26 per cent in 2005/06.*

Recycled household materials

Materials collected from household sources for recycling: 2006/07, England

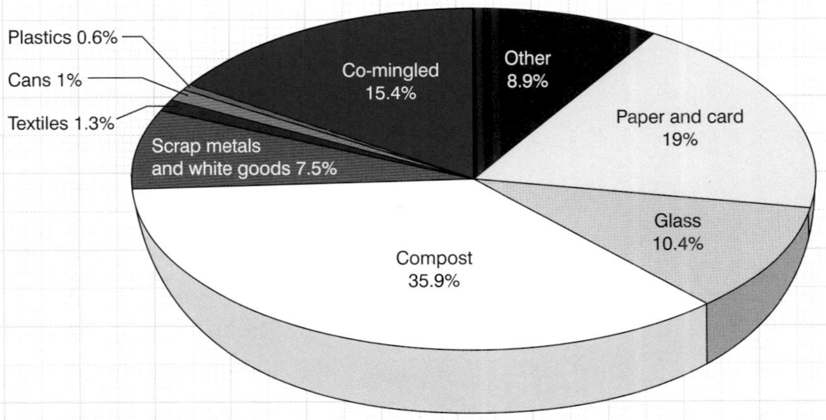

Source: Defra. Crown copyright.

Mini glossary

convenient – *easy to reach*

contaminated – *made impure by contact with dangerous or unhealthy substances*

pilot scheme – *a trial of a policy to test how effective it is*

www.edie.net

Why recycling isn't really saving the planet

THE GOVERNMENT CONSTANTLY TELLS US that recycling is a wonderful idea. TV advertisements bombard us with this message. Children are taught to recycle at school. But are the environmental benefits of recycling really worth the economic cost?

Cost

The cost of recycling is growing rapidly.

- The landfill tax, introduced to encourage recycling, increased by 14 per cent in April. This cost UK businesses and householders £900m in 2007.
- The costs of collecting and reprocessing the recyclable materials amount to an estimated £400m per year, paid for by council tax payers.
- A number of EU waste directives are also being forced on different industries. The Waste Electronic and Electrical Equipment Directive, for example, cost British business £200m in 2006.
- Despite all this expenditure, running into billions, it is difficult to identify any significant benefits from recycling – except where it takes place commercially without government involvement, as in the traditional scrap metal business.

Landfill myths

The amount of municipal waste being sent to landfill has gone down, but then again modern landfill sites have few negative environmental impacts.

- They are lined so that pollution cannot seep into water courses. Unpleasant odours and dust are carefully controlled.
- Often landfill has a positive role in rehabilitating landscapes scarred by quarrying, such as the Yorkshire Wolds.
- It is a myth that there is no more room for landfill. Currently, landfill sites take up a little over one-tenth of one per cent of the UK. Since rubbish rots and can be piled up, there is no practical limit on the amount of waste that can be stored. Any shortage of landfill space has been artificially created by the planning system and a government desperate to provide a reason for costly recycling policies.

Why recycling isn't really saving the planet

Environmental arguments

Another argument is that recycling saves energy and therefore reduces carbon emissions. Yet the amount saved is trivial, and there are far more efficient ways of reducing consumption.

- Even if every steel can in the UK was recycled the amount saved would amount to less than one-thousandth of the country's annual energy usage.
- Many people also believe that recycling paper saves trees. In fact, the opposite is true. Recycling reduces the demand for wood pulp and reduces the value of trees, so fewer are planted. This is a shame because young forests are particularly good at absorbing greenhouse gases.

Economic growth

Recycling may also be having a damaging effect on many Third World countries.

- Some, such as Jamaica, are heavily reliant on exports of raw materials such as bauxite (to make aluminium) and iron ore (to make steel). By subsidising recycling, the Government is effectively reducing the demand for imported raw materials while artificially supporting the domestic waste-processing industry.
- Recycling both reduces the income of poor countries that rely on raw material exports and reduces economic growth in industrialised countries burdened with additional taxation and red tape.

Mini glossary

expenditure *– amount of money spent*

rehabilitating *– restoring land to a good condition*

quarrying *– extracting stone from a mine*

trivial *– small and insignificant*

subsidising *– supporting financially*

exports *– goods sold to other countries*

red tape *– slow or unnecessary government procedures*

misguided *– badly thought out*

Promoting recycling

Given that the environmental case is weak and that the economic case is non-existent, why does the Government continue to promote recycling?

- One answer is that it doesn't have much choice. EU legislation means that unless the amount of household waste recycled is increased dramatically, local councils will face huge fines of up to £150 per tonne.
- Within local authorities there are also those with much to gain from the growth in recycling. A growing army of officials has been employed to supervise the carrying out of EU targets by microchipping wheelie bins and rummaging through other people's rubbish.
- Many waste-processing companies have also gained from recent legislation, and these make up an influential group in favour of further protection and subsidy for their industry.

Losing out?

The main losers are ordinary householders facing increases in council tax, higher prices in the shops and being forced to perform the pointless chore of separating out their tin cans and newspapers.

With the introduction of heavy fines for those who refuse to recycle, it remains to be seen how much state control will be necessary to enforce these costly and misguided policies.

***By Richard Wellings, Deputy Editorial Director** at the Institute of Economic Affairs*

www.iea.org.uk

Consumers oppose charges for plastic bags

NEW RESEARCH FROM Ipsos MORI indicates that the government may face consumer opposition to possible legislation which will force retailers to charge for plastic bags – particularly from less affluent shoppers who resent being dictated to on this issue.

Making up their own minds

Although 57% of the population agrees it is right for retailers to start charging for plastic shopping bags, 52% of people feel they should be able to make up their own minds about this issue and not have it forced on them by retailers wanting to charge. 35% feel it is wrong for the government to ban retailers from giving them away free.

Significantly it is Londoners who are most against having to pay for plastic shopping bags. Also, of the 22% who agree that they would prefer to pay for a bag rather than bring their own, a higher proportion than average are young shoppers aged 15-24 and affluent shoppers.

Comments Gill Aitchison, President, Global Shopper & Retail Research, Ipsos MORI:

“Our findings indicate that although people pretty much accept the concept of being charged for carrier bags, they don't appreciate being forced to accept this either by retailers or the government. Unsurprisingly, the move is least popular with less affluent consumers as currently they often use carrier bags as substitute bin bags – while for Londoners the idea of having to take a shopping bag with you when you go shopping can present a practical problem.”

Convenience

“It is interesting to note the attitudes of the young and wealthy on this topic. While conventional thinking might suggest this group is more concerned with protecting the environment, they are not apparently willing to sacrifice the convenience of plastic bags and would rather pay than bring their own when they go shopping.”

Proceeds of funds

When it comes to the proceeds of funds raised through charging for carrier bags, views of the public seem very clear. 74% of the population feel it is right for money raised to go to charity – this rises to 84% for 15-24 year olds and is shared even by those who disagree with the principle of charging for bags (76%)

Re-usable bags

Concludes Gill Aitchison:

“In addition, given the relatively high number of respondents who said that they would rather bring their own bag than pay for one (59%), there are signs that the Anya Hindmarch re-usable shopping bag trend is set to intensify”

13 March 2008

Mini glossary

opposition – *resistance to something you don't agree with*

affluent – *financially well-off*

conventional – *usual*

www.ipsos-mori.com

Shipping our plastic bottles to China

A NEW WRAP STUDY has found that selling the UK's used plastic bottles and paper for recycling in China actually saves carbon emissions. Shipping these materials more than 10,000 miles produces less CO^2 than sending them to landfill at home and using brand new materials.

The transport issue is just one factor in assessing the environmental impact of exporting materials for recycling. However, it has become increasingly important to understand, as over the last ten years exports of used paper have risen from 470,000 tonnes to 4.7 million tonnes. Exports of used plastic bottles have gone from less than 40,000 tonnes to half a million tonnes over the same period.

CO^2 emissions

The study showed that the emissions caused by transporting the material to China account for only a small amount – on average less than a third – of the CO^2 saved by recycling. However, due to the imbalance of trade between China and the UK, most container ships head back to China empty and they are producing CO^2 emissions whether or not they are carrying goods. If you take this into account, the transport emissions are even smaller – less than one-tenth of the overall amount of CO^2 saved by recycling.

Rise in recycling

This increase reflects the huge rise in household recycling in the UK from 7% to over 30% during that time. We collect more paper than we can recycle, but there is strong demand for it from growing economies, such as China, where there are not enough trees to make paper.

Plastic bottles are also much in demand from China's manufacturing industry and there is currently not enough capacity in the UK to recycle them here. This study shows it is environmentally less harmful to send that material to China for reprocessing than sending it to landfill in the UK.

Liz Goodwin, CEO, WRAP, said:

“It may seem strange that transporting our unwanted paper and plastic bottles such a distance would actually be better for the environment but that is what the evidence from this study shows.

Although this study is only part of the environmental impact story, it is clear that there are significant CO^2 savings that can be made by shipping our unwanted paper and plastic to China. In some cases, we just aren't able to reprocess everything we collect or there isn't enough of it to do so. In these cases, shipping it to China, which has a high demand and need for material, makes sense in CO^2 terms.

WRAP will continue to build both the environmental and economic case for domestic recycling.”

Mini glossary

capacity *– amount of material that can be dealt with*

domestic *– within one's own country*

19 August 2008

The above information is reprinted with kind permission from WRAP.

www.wrap.org.uk

Activities

Brainstorm

Brainstorm to find out what you know about recycling.

1. What are the 'three Rs'?

....

....

2. What is recycling?

....

....

3. What is composting?

....

....

Oral activities

4. In a group, discuss how much your families recycle and any steps you have personally taken to reduce your household waste.

5. With a partner, prepare and record an interview as if for radio in which one of you argues the benefits of recycling and the other argues against it. Use the articles *Waste at home* on page 13 and *Why recycling isn't really saving the planet* on pages 22-23 as a starting point.

NOTES..

....

....

....

Moral dilemmas

6. Imagine a new landfill site is going to be built near to your home. The rubbish we all create has to go somewhere, but you are worried about the noise, smells and pollution the landfill site may cause. Would you protest against the site? How do you think this would affect your views on waste and recycling?

7. The government is considering charging households for how much rubbish they throw away in order to cut down on the amount of waste sent to landfill. Imagine you are an MP voting on this issue in Parliament. How do you vote? Give your reasons.

Activities

Research activities

8. Do some research on recycling in your school. How many recycling bins are there? Which materials can you recycle at school?

NOTES ....

....

....

....

9. Visit the Recycle Now website at www.recyclenow.com and write a brief summary of how the following materials can be recycled:

DRINKS CANS ....

....

COMPUTERS ....

....

MAGAZINES ....

....

TEXTILES ....

....

Written activities

Complete the following activities in your exercise books or on a sheet of paper.

10. Create a checklist of five simple things everybody could do to reduce waste.

11. Look at the quiz *How much of a waster are you?* on pages 16-17. Devise a similar quiz focussing on reducing waste in the home. You will need to think about the kind of waste that is produced at home and come up with a range of options for dealing with this waste.

Design activities

12. The government wants us to recycle on the go as well as at home. Design a poster that could be placed in public areas to encourage people to recycle when they are out and about.

Key Facts

- *In 2005/06 on average each person in the UK produced 511 kg of waste, 74 per cent of which was not recycled. (pages 1-2)*
- *The rubbish from businesses and industries makes up the largest proportion of rubbish collected in the UK. (pages 1-2)*
- *In 2004 the UK produced about 335 million tonnes of rubbish. This rate of rubbish production would fill the Albert Hall in London in less than two hours. (page 3)*
- *64% of the UK's municipal waste was sent to landfill in 2005/06. 27% was recycled or composted, and 8% was incinerated. (pages 4-5)*
- *In the UK we are throwing away one-third of the food we buy. That's like one in three bagfuls of food shopping going straight in the bin. (page 6)*
- *Up to 38 per cent of packaging in a regular household shopping basket cannot be recycled. (page 7)*
- *It is illegal to drop litter and you can be fined up to £80 on the spot if you're caught littering. (page 8)*
- *Over 69,000 animals were killed or injured by litter last year in Britain. (page 8)*
- *More than 13 billion carrier bags are distributed in the UK every year. (page 9)*
- *When we bury or burn our rubbish we are losing valuable natural resources and wasting the energy, water and transport costs used in its production. Landfilling and incineration can harm the environment if they are not properly managed. (page 13)*
- *The unreleased energy contained in the average dustbin each year could power a television for 5,000 hours. (page 14)*
- *The UK currently recycles around 50% of container glass (like bottles and jars). That's doubled over the last 5 years. (page 20)*
- *The amount of household waste recycled or composted in the UK increased from 6 per cent in 1995/96 to 26 per cent in 2005/06. (page 21)*
- *Currently, landfill sites take up a little over one-tenth of one per cent of the United Kingdom, Richard Wellings of the Institute of Economic Affairs has said. (pages 22-23)*
- *Although 57% of the population agrees it is right for retailers to start charging for plastic shopping bags, 52% of people feel they should be able to make up their own minds about this issue and not have it forced on them by retailers wanting to charge. 35% feel it is wrong for the government to ban retailers from giving them away free. (page 24)*

Glossary

Biodegradable – *Materials that can be broken down by air, water and bacteria, and so have less negative environmental impact than non-biodegradeable materials.*

Carbon emissions – *Environmentally harmful gases given off.*

Composting – *The natural breakdown of organic waste by the bacteria, insects and animals in the soil.*

Consumption – *Buying and using goods and services.*

EU Landfill Directive – *The directive requires EU member states (which include the UK) to decrease the amount of biodegradable material they send to landfill to 35% of 1995 levels by 2016.*

Energy from waste – *Energy from waste or 'waste to energy' refers to the process of burning waste in order to generate electricity.*

Fly-tipping – *Term used to describe illegally dumping waste in an area of land that is not an authorised landfill site.*

Greenhouse gas – *A gas, such as methane, which traps heat in the atmosphere and contributes to climate change.*

Hazardous waste – *Any waste that could be a serious danger to humans or other living organisms, such as chemicals, clinical waste or fuel.*

Household Waste Recycling Act 2003 – *This act set out a target for all English waste collection authorities to offer households in their area a doorstep collection of at least two types of recyclable waste by 2010. 90% of households in England currently have this.*

Incineration – *A method of disposing of waste by burning it.*

Landfill – *At a landfill site, waste is deposited into holes in the ground, which are then covered over.*

Landfill tax – *A tax charged on waste sent for disposal at a landfill site, in order to discourage waste production and encourage environmentally-friendly methods of disposal.*

Litter – *Rubbish that has been dropped in a public place rather than disposed of properly in a bin.*

Municipal waste – *Municipal waste refers to all household, commercial and industrial waste.*

Natural resource – *Any material found in the natural enivornment and used by people.*

Pollute – *Make harmful to living things.*

Recycling – *The process of turning something considered waste into a new product. Recycling used materials reduces consumption of natural resources, saves energy and reduces the amount of waste sent to landfill.*

Sustainable – *Capable of being continued with little long-term effect on the environment.*

The three Rs – *The three Rs of rubbish are to Reduce, Reuse and Recycle (sometimes called the waste hierarchy).*

Waste – *Anything that is no longer of use and is thrown away.*

Waste Electrical and Electronic Equipment (WEEE) directive – *This directive makes the manufacturers of electrical goods responsible for financing the collection, treatment, recovery and environmentally-sound disposal of those goods.*

Zero waste – *A situation in which all waste is either reduced, re-used or recycled, so there is nothing left of which to dispose.*